···· In · Time · of · Need ····

Fire

by Sean Connolly

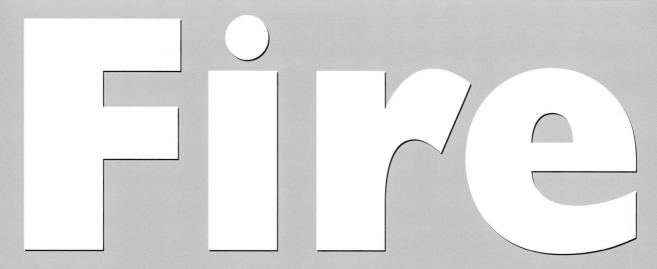

A⁺
Smart Apple Media

Published by Smart Apple Media
2140 Howard Drive West
North Mankato, Minnesota 56003

Design by Ian Butterworth

Photographs by:
Corbis (Daniel A. Anderson/Orange County Register,
Mark Avery/Orange County Register, Dana R. Bowler/Ventura
County Star, Bruce Chambers/Orange County Register,
Michael Goulding/Orange County Register, Steve Hix,
Karen Quincy Loberg/Ventura County Star, Matt McClain/Ventura
County Star, NASA, Lucy Nicholson/Reuters, Charles O'Rear,
Royalty-Free, Bill Stormont, David Turnley, Russell Underwood)

Library of Congress Cataloging-in-Publication Data

Connolly, Sean, 1956–
Fire / by Sean Connolly.
v. cm. — (In time of need)
Includes index.
Summary: Describes fire, how it is both useful and dangerous, and
what can be done to protect people, animals, and the environment
from all sorts of fires.
Contents: A dangerous tool—Types of fire—Among the flames—
Fighting fires—The damage done—Working together—
The battle continues.
ISBN 1-58340-392-2
1. Fire extinction—Juvenile literature. 2. Fire—Juvenile literature. [1.
Fire extinction. 2. Fire.] I. Title.

TH9148.C596 2003
628.9'2—dc21
2003042488

First Edition

2 4 6 8 9 7 5 3 1

Contents

A Dangerous Tool

We all depend on fire in many ways. It can produce heat to cook our meals and warm our homes. A blazing fire welcomes us on a cold winter's day. Human beings have used fire in these and many other useful ways for thousands of years. Many scientists consider the use of fire to be the first human invention. Early humans used fire to scare off dangerous animals and warm caves and campsites.

But even these first users of fire knew that it also had its dangers. Fire can injure and kill people, and it can destroy the very homes that it heats. If it is not put out in time, a fire can quickly spread and destroy a much larger area. Knowing how to control fire is as important as

Above: When contained and controlled, fire is an invaluable tool. Left: But if used carelessly, fire can be terribly destructive.

knowing how to harness its heat. Careless burning—in the home or outside—can lead to a deadly fire. Television news broadcasts show houses, schools, and even entire forests burning almost every evening. When such fires blaze up, people need to know how to find safety, raise an alarm, and stop the fire from spreading. But most of all, people need to know how to prevent dangerous fires from starting.

Tackling the Problem

Human beings have tried to protect themselves from fire since their earliest days. When people began living in towns and cities about 5,000 years ago, the danger of fire increased. Every home used open fires for cooking and heating, and people lived close to each other. Most houses were built of wood or other materials that burn easily. A small fire in one household could spread quickly throughout a town.

Below: About 80 percent of all fire-related deaths occur in residences, and it is estimated that more than half of these fatalities occur in homes without smoke alarms.

Below: It can take less than 30 seconds for a small flame to become a major fire—leaving no time for people to grab personal belongings before heading to safety. Opposite: Fighting fires is a dangerous but important job.

The first steps to deal with dangerous fires were taken in ancient Rome more than 2,000 years ago. The Roman emperor Augustus created a group of *vigiles* ("watchmen") in 24 B.C. These watchmen did not try to put out fires. They simply sounded an alarm to warn other people about the danger. The job of firefighting as we know it came much later, after the Great Fire of London in 1666. **Insurance** companies, which had to pay for fire damage, saw that they could save lives—and money—by paying fire **brigades** to put out fires. The British government formed the first modern firefighting service in 1865 by combining these fire brigades into London's Metropolitan Fire Brigade. Soon, other cities and towns around the world had firefighting services prepared to tackle fires at any time of the day or night.

Whether it begins in a town or a forest, a fire is dangerous for those living near it and those who come to put it out. Families and companies still use insurance companies to pay for damage caused by fires. But no amount of money can replace a family member killed in a fire—or even a treasured toy or photo album. The wider world also suffers when a fire destroys a much-loved building or a stretch of wild forest.

WHAT IS A FIRE?

A fire is a REACTION between a FLAMMABLE substance and a gas called OXYGEN. The reaction is called COMBUSTION, and it is accompanied by heat, light, and sometimes an open flame.

There is oxygen in the air all around us. It can create a slow combustion—also known as rusting—when it reacts with certain metals. This reaction gives off heat, but it happens so slowly that people do not notice it. If a flammable substance warms up to a certain temperature, it can react with the oxygen to produce a quicker combustion: a fire. We can see and feel the light and heat produced by this reaction. The burning material is broken down into gases and small, floating bits of material—this is called smoke. The warm gases are visible as flames. Some materials, such as paper, gasoline, and the sulfur on matches, are very flammable. Others, such as the materials used for clothing, are far less flammable.

Combustion continues until either the material burns away or the oxygen runs out. If you put a glass jar upside-down over a burning candle, you can see the flame die down and then go out. The fire is EXTINGUISHED because it has used up all the oxygen inside the jar.

According to the World Wide Fund for Nature, more than 12 million acres (4.8 million ha) of forest around the world burned in 1997 alone. This is an area about the size of Vermont and New Hampshire combined.

Types of Fire

The two main types of fire are structure fires and **wildfires**. Structure fires are those that affect houses, offices, and other buildings. Many such fires develop because of carelessness or accidents. For example, splattering oil from a frying pan can easily start a fire if it lands on the open flame of a gas stove. Some electrical products such as irons or hairdryers can start fires if they are left too close to cloth, paper, or other materials that may catch fire quickly. Cigarettes, candles, and other burning objects pose even more of a risk. Fires in larger buildings often have similar causes. One careless action in a hotel, office building, or school can lead to a damaging fire—and great tragedy. The Great Fire of London destroyed four-fifths of the city, including 13,200 houses and 87 churches, in 1666. It started when a baker left an oven burning too close to some woodchips after he went to bed.

Above: The room temperature of a fire can be 100 °F (37 °C) at the floor and rise up to 600 °F (315 °C) at eye level. Left: The suffocating smoke of a wildfire can make it hard for firefighters to breathe.

Wildfires

During the summer of 2002, a series of terrible fires burned out of control for weeks in the western United States. Flames swept through forests that had gone for weeks without rain. We normally call these forest fires, but fire experts prefer to describe them as "wildfires" or "wildland fires" because some of them burn through areas that are not really forests. For example, fires can just as easily sweep across prairies, small woods, and even swamps and other areas that are normally damp. There is a natural need for wildfires, to clear away dead wood and generate new growth, but they can easily spread out of control and endanger human life and property.

The heat produced by many burning trees causes the air to rise rapidly, sucking in more air. The wind rushing in can reach speeds of more than 120 miles (195 km) per hour, causing flames to leap from treetop to treetop. The flames may be as high as 300 feet (91 m), and the smoke may rise to heights of 40,000 feet (12,200 m).

Below: The sky is filled with billowing smoke as wildfires rage in southern California. In 2003, at least 2,612 homes were destroyed by wildfires that left 20 people dead and more than 950,000 acres (385,000 ha) scorched.

WHEN FIRE IS A CRIME

Deliberately starting a fire is a serious crime known as ARSON, and the crime can become murder if people die in the fire. At least half of Australia's damaging wildfires in 2001 and 2002 were thought to be started deliberately. Australian police arrested 24 people in connection with the fires, and 15 of those arrested were between the ages of 9 and 16. Local government officials decided to require all children found guilty of arson to visit hospital patients suffering from severe burns so they can see the effects of their crimes.

Dangerous Mixtures

Some of the most dangerous—and hardest to extinguish—fires occur when electrical or chemical equipment is burning. Unlike fires that burn wood, cloth, or similar materials, these fires cannot always be put out with water or simply smothered with a **fire blanket**. An electrical fire, for example, is difficult to put out because water **conducts** electricity. Firefighters must use special chemicals that do not conduct electricity to put out electrical fires.

Oil, gasoline, and similar substances also call for special equipment when they catch fire. In July 2000, firefighters battled a terrible fire in Nigeria, a country in west Africa. Thieves had broken a stretch of pipeline in order to steal gasoline, but their tools caused the whole pipeline to catch fire. More than 250 people died in the blaze. Firefighters needed to use a special foam to put out the fire—water alone would not have worked in the extreme heat.

Above: Because water can cause electrical fires to spread, firefighters must use foam and chemicals to extinguish the flames.

SLASH AND BURN

People living in areas near tropical forests or jungles often use fire to clear new land for planting crops. This method is called "slash and burn." First they cut or "slash" down trees, bushes, and shrubs. Then they burn all the fallen wood. After the fire has burned down, the field is covered in ash, which makes the soil better for growing many crops. Slash-and-burn farming is very destructive to the environment. It destroys parts of forests—and with them the homes of many animals. Trees also produce oxygen, which animals (including humans) need to breathe. Sometimes slash-and-burn fires get out of control, and farmers have little chance of putting out a large forest fire once it starts.

Left: Burning trees to create farmland destroys not only forests, but animal habitats as well.

In wildland areas, 80 percent of all fires are caused by lightning. Of the eight million lightning strikes that occur worldwide each day, 80,000 create wildfires.

Among the
Flames

Outside a burning building, firefighters often put up barriers to keep people from taking dangerous risks just to get a better view of the fire. People watching a fire are following a basic urge to watch a drama unfold, just as a crowd might gather around two people fighting. They will go home with stories to tell. But the stories of those who are inside the fire—the trapped victims and the firefighters working through smoke, flames, and falling buildings to reach them—are very different. They are the real accounts of a fire, full of fear and respect for one of nature's deadliest weapons.

Above: Although it may be tempting to see a fire up-close, it is important to let firefighters do their job and to respect a fire's power and unpredictability.

Some of these stories demonstrate that survival may depend on luck—stumbling upon a way out or being able to break a window to get enough air to carry on. Other stories tell of the courage of firefighters and other rescuers who save lives in the most dangerous conditions. Still others—usually told by those who search through the burned remains of a fire—are sad accounts of victims who collapsed just a few steps away from fresh air and safety. Every story points to the simple fact that fire is a killer that can strike suddenly over a wide area.

SWEDISH TRAGEDY

More than 60 people, mainly teenagers, died on October 31, 1998, in what became the deadliest fire in Swedish history. The fire swept through a nightclub in Gothenburg, Sweden's second-largest city. About 400 young people were attending a Halloween dance, but the hall was licensed to hold only 150 people. The fire broke out at midnight in an upstairs room where the dance was being held. A blocked fire exit trapped many victims inside. Fifteen-year-old Emil Fawn, who was in the room, said the fire seemed to have started from the ceiling as "lamps and loudspeakers fell to the floor. It was chaos. Everybody was trying to get out and people trampled on each other on the way to the exit. . . . Others kicked out the windows and jumped out."

Below: Firefighters may need to flee at a moment's notice to escape sudden wildfire flare-ups.

JUST SIX FEET FROM SAFETY

One morning in August 2002, Keith Filcher of Houston, Texas, was on his front porch when he noticed white smoke rising from a building about 100 feet (30 m) away. Filcher, a firefighter for more than 20 years, knew that the smoke spelled trouble and ran toward the burning first floor apartment. "I was fully involved in, I'd say, three to four minutes," he recalled later. Filcher and another witness, Sean Burrows, broke down the front door.

"We yelled, 'Is anyone in there? Is anyone in there?' And then she yelled, 'Yes, yes!' I couldn't really understand a word that she was saying, but you could tell that there was someone in there. We tried to go, we tried to make entry on it, but we were on the ground," said Filcher. "The smoke was just two or three feet off the ground. It was black smoke. We couldn't go in. It was too thick."

Filcher and Burrows then rushed to the apartments upstairs. They helped to EVACUATE more than a dozen people by the time the fire engines reached the fire. When firefighters arrived, they found the body of a 24-year-old woman lying only six feet (1.8 m) inside the front door.

"Wounded bears have been wandering into town and I have seen waterfowl and eagles die from the heat and smoke and fall from the sky in mid-flight."
Eyewitness Scott Piney describing the 2002 forest fires in Colorado

"Last Thursday when flying back to Arizona from Denver, we flew right over the Show Low area. There was black ash in the clouds up to 20,000 feet (6,100 m), and looking down on one of the **hot spots** was like looking into the mouth of a volcano."
Air passenger Phil Briers describing the worst hit area of the 2002 Arizona wildfire

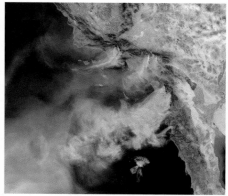

Above: A photo of the 2003 southern California wildfires taken from space.

"I live and work in the southwestern corner of the Denver metro area, **downwind** of Colorado's Harman fire. In the first few days of the fire, the smoke was pretty awful. Even indoors, there was no way to get away from it. I could look out the window of my apartment and see ash falling."
Scott Piney describing conditions during the 2002 Colorado wildfire

Fighting
Fires

The main firefighting tool in ancient Rome and as recently as the 19th century was the bucket, passed from hand to hand to deliver water to the fire. Early firefighters used axes to remove flammable material such as wood or cloth from a fire. Axes were also used to make openings to allow heat and smoke to escape from a burning building.

Today's firefighting teams use some similar methods, although modern inventions have made it easier to locate fires and battle the flames. Using a combination of tele-phones, cellular phones, and computers, fire departments send crews to the scene of the fire. Then, working closely as a team, they spring into action.

Above: Today's firefighters undergo extensive training to prepare them for unexpected and difficult fire situations. Left: When roads are accessible, fire trucks take firefighters where they are needed most.

On the Scene

The first job of firefighters is to look for and rescue people who are trapped inside. Then fire crews turn their attention to protecting other buildings exposed to the fire and limiting the fire's spread. Taking these steps first makes it easier to put out the fire itself.

Modern fire engines draw water through hoses connected to **hydrants**. A powerful pump on the fire engine draws water from the hydrant and sends it through thousands of feet (hundreds of meters) of hose so that firefighters can reach the highest flames. Some fire engines must carry their own supply of water if there is no hydrant near the fire. Sometimes helicopters hover over a fire, ready to release large amounts of water onto the highest parts of city buildings. Eventually, the fire is controlled and finally put out. But the job is not yet over. Some firefighters remain on the scene to make sure that the fire does not start to burn again.

Below: Air tankers drop fire retardant on hot spots to keep wildfires from advancing into local communities or nearby homes.

VOLUNTEERS

Most large towns and cities in the U.S., Canada, and other countries have PROFESSIONAL fire departments to put out fires. In smaller communities and in most rural areas, VOLUNTEERS are called in to battle fires. In the U.S., there are nearly one million volunteer firefighters. In Canada, there are more than 120,000 volunteers, making up 80 percent of Canada's overall firefighting force.

Other countries with large areas of wilderness also rely on volunteers to fight fires. Nearly half of the 15,000 firefighters who battled Australia's wildfires in 2001 and 2002 were volunteers. Russia, Finland, and other forested countries have similar forces ready to go to the scene of wildfires.

PUTTING OUT FIRES

All fires need oxygen to burn. A fire will stop burning if the burning material cannot get more oxygen—because it is being smothered by a fire blanket, for example. Cooling the material, as with water from a fire hose, can also put out the fire. These simple facts lie behind all the different ways of putting out fires. Water or a fire blanket will be enough to put out a small fire of burning wood. Larger amounts of water—either pumped from a hose or dropped from the air—help put out larger fires in buildings and forests. Some fires, however, call for special materials. For example, firefighters use foam or special chemicals to smother chemical fires because the temperature is too high for water to be effective.

In the Wild

Although wildfires are much larger than structure fires, the people tackling them use a similar approach. Once again, rescuing people is the first job. Then firefighters must make sure of the exact size of the fire and the direction that the wind is blowing the flames. It can be hard to put out a wildfire directly because of its size and the extreme heat near it, so firefighting teams instead work to control it. One of the best ways to do this is to create a gap in the woods called a **firebreak**. If the firebreak is wide enough, the flames from the burning trees will not reach across to the other side. The sides of the firebreaks are soaked with water or chemicals to make it harder for the fire to pass through. Planes and helicopters drop water along the edge of the fire. Some brave fighters known as **smokejumpers** actually drop into a fire zone by parachute to help plug any gaps in the firebreak.

"The high school where my father taught and my sisters and I attended is now a staging area for men with **pickaxes**, walkie-talkies, . . . and **fire-resistant** clothing."
Eyewitness Matthew Francis describing Arizona's forest fires of June 2002

Opposite: A firefighter lights backfires, hoping to create a firebreak to halt the oncoming blaze. Below: There are between 15 and 20 helicopters in the U.S. used only for fighting wildfires, with another 175 available for use if needed.

America's first volunteer fire department was founded in Philadelphia, Pennsylvania, after a serious fire in 1736. The famous statesman and inventor Benjamin Franklin created a fire brigade called the Union Fire Company with 30 volunteers. Other cities followed Franklin's example. Famous Americans who served as volunteer firefighters included George Washington, Thomas Jefferson, and Paul Revere.

The Damage

No matter how they start, fires can cause terrible damage by the time they are put out. The worst damage, of course, is to human life. People can be burned to death in the terrible heat. More people die from breathing smoke, which causes them to **suffocate**. Many people are hurt or killed by falling walls or roofs. About two million fires occur each year in the U.S. alone. These fires result in about 5,000 deaths and 1.4 million injuries. Worldwide, there are as many as 24 million fires each year, leading to 160,000 deaths. Fire experts estimate that the overall cost of fires around the world is as much as $400 billion a year.

Unfortunately, children are at the heart of many fires—either starting them or suffering from them. American children under the age of 10 with access to cigarette lighters and matches cause about 100,000 fires, 300 to 400 deaths, and 11 percent of all reported fire injuries each year.

Above: More than 30 percent of the fires that kill children are actually started by children playing with fire. Left: Fire-proof safes protect money and other valuables.

Sudden Destruction

A fire can also cause a great amount of damage to land and property. Tall buildings, schools, and private homes can be turned into smoking shells in just a few hours. Even the buildings still standing might have hidden damage. Flames can cause destruction behind walls and above ceilings, ruining the insulation, electrical wires, and other important features. The water used to put out the fire can ruin carpets and furniture and cause other damage.

Wildfires can create even more destruction. In the first 10 days of Australia's Christmas 2001 fires, an area about the size of Delaware was burned and more than 170 homes were destroyed. The bill for that damage came to more than $36 million.

A DEADLY ACCIDENT

The worst industrial disaster in U.S. history was a factory explosion in Texas City, Texas, in 1947. The explosion was most likely caused by a cigarette. The blast caused nearly 600 deaths, injured more than 4,000 people, and damaged more than 90 percent of the city's buildings. Overall, the explosion and fire caused more than $4 billion in damage.

PUBLIC EDUCATION

One of the best ways of preventing—or at least reducing—the damage caused by fires is to educate the public about fire and fire safety. Many fire departments around the world send representatives to schools and community centers to provide basic information on fire prevention and safety precautions in the event of a fire. For example, the Volunteer Fire Department in Gravenhurst, Ontario, keeps the local population aware of many aspects about fire safety; visits from fire officials and fire-safety competitions provide public education about fire alarms, fire prevention, escape drills, arson, and forest fires.

Left: In less than five minutes after a room catches on fire, the temperature can get so hot that everything in it ignites at once—engulfing the room in flames.

Working Together

Every house, school, office, and public building should have fire alarms to alert people to fires, as well as equipment to put out small fires before they spread. There should be several ways of leaving every building in case any one of them becomes blocked by fire or thick smoke. Even small children should know where to go at the first sign of a fire. Basic safety depends on this combination of equipment and fire awareness. It is important that everyone is aware of fire safety. Really dangerous fires may spread because firefighters are kept busy by smaller fires that could have been prevented or easily extinguished. Everyone has a part to play in fighting fires.

Sometimes, fire departments in a community need extra help because a fire has grown so big that they cannot put it out on their own. At first, they call out more of their own firefighters and equipment. American fire departments

Above and left: Firefighters often work together with other local fire departments when fires get too large for one department to handle. Sometimes state, national, and even international assistance is needed—especially for wildfires.

describe this as the "alarm" system. A "three-alarm fire" means that two fire-fighting units in addition to the first team are needed to fight a fire. Even this might not be enough, and a fire department may call on neighboring communities to supply more firefighters and equipment.

Fighting very large fires—especially wildfires—often requires state, national, and even international help. In the U.S., a state governor or the president may decide that a particular fire has become a major emergency in need of special attention. This makes additional money available for use in fighting the fire, providing emergency aid to those affected by the fire, and rebuilding. The **National Guard** may be called out to help combat fires that have become widespread. Other countries can mobilize fire departments from other regions in an emergency. Japan operates 43 regional and three city fire brigades, any of which can be called on to help outside their region. France has a similar system, backed up by special brigades operated by the army in Paris and by the navy in Marseilles.

ELVIS TO THE RESCUE

The U.S. National Guard sent several HELITANKERS to help Australian fire services battle wildfires in late 2001. These helicopters can lift 28,000 pounds (12,700 kg) and release up to 2,500 gallons (9,500 l) of water in one drop. The Australians nicknamed one of them "Elvis" because it came from Memphis, Tennessee, where singer Elvis Presley became famous. While dumping water on the wildfire, the pilot of Elvis responded to an emergency call from 14 firefighters who were trapped behind a wall of flame. The helitanker was able to carry the firefighters back to safety.

Below: A Los Angeles City Fire Department helicopter drops water on a fire to prevent it from jumping a freeway in southern California.

Below: The melted remains of a television news van are all that is left after the vehicle was engulfed by flames.

Help from Other Countries

Countries sometimes turn to each other for help in fighting particularly bad fires. This sort of cooperation is common in Europe, where there are many countries near each other. In March 1999, for example, firefighters from France, Italy, and Switzerland battled a fire that had developed in France's Mont Blanc road tunnel. Fire experts believe that the fire began when a driver carelessly tossed a lit cigarette from a car window. Working around the clock in temperatures twice as hot as an oven, the international team rescued dozens of passengers who had been trapped. One Italian rescuer, Pierlucio Tinazzi, was one of 39 people who died in the fire. He had saved 10 people by making repeated trips into the tunnel on his motorcycle before he was killed.

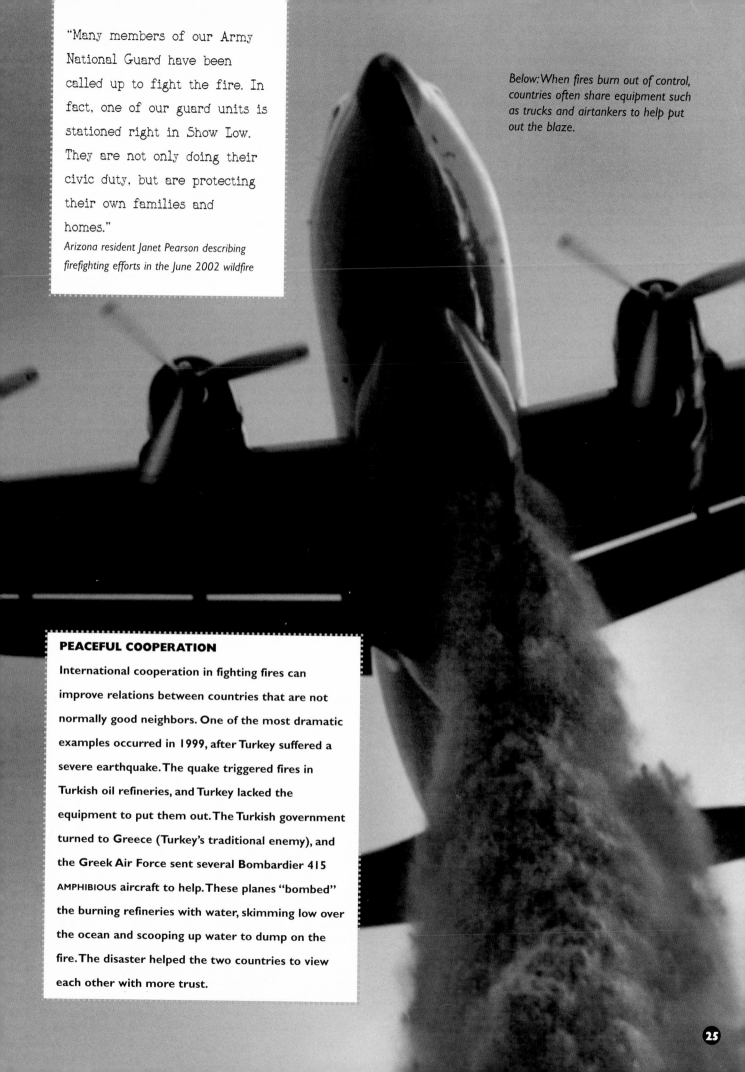

"Many members of our Army National Guard have been called up to fight the fire. In fact, one of our guard units is stationed right in Show Low. They are not only doing their civic duty, but are protecting their own families and homes."

Arizona resident Janet Pearson describing firefighting efforts in the June 2002 wildfire

Below: When fires burn out of control, countries often share equipment such as trucks and airtankers to help put out the blaze.

PEACEFUL COOPERATION

International cooperation in fighting fires can improve relations between countries that are not normally good neighbors. One of the most dramatic examples occurred in 1999, after Turkey suffered a severe earthquake. The quake triggered fires in Turkish oil refineries, and Turkey lacked the equipment to put them out. The Turkish government turned to Greece (Turkey's traditional enemy), and the Greek Air Force sent several Bombardier 415 AMPHIBIOUS aircraft to help. These planes "bombed" the burning refineries with water, skimming low over the ocean and scooping up water to dump on the fire. The disaster helped the two countries to view each other with more trust.

The Battle Continues

Fighting fires has changed a great deal since the first caveman dumped water on a fire and ancient Romans patrolled their crowded cities. Satellites and high-flying aircraft can spot likely fire spots or provide early warning when fires break out in remote areas. Emergency 911 calls trigger automatic alerts to fire departments that inform them of the location and size of fires. Modern communications equipment gives firefighters the information they need to send the right number of firefighters and the right tools to tackle nearly every type of blaze.

At the scene of the fire, planes and helicopters check the fire's progress from overhead, sometimes dropping water or chemicals on parts of a fire that cannot be reached from the ground. Special foam, dropped from above or shot from a firefighter's hose, can also smother flames. Firefighters wear protective suits that enable them to go into areas that are very hot or choked with dense smoke or poisonous chemicals.

Above and left: Even with modern equipment and help from helicopters and airtankers, about 100 U.S. firefighters are killed each year in duty-related incidents.

Looking Ahead

All of this comes at a price. Keeping firefighting equipment in good condition costs money, and fire departments might be stretched if they have to battle two or more huge fires at the same time. Likewise, a state's or region's fire services can be stretched to the limit fighting a large-scale wildfire. One solution to the problem is obvious—trying to prevent fires before they start.

Robert Henricks, an official in the U.S. Department of Agriculture Forest Service, says that one of the most important methods of fire prevention is educating the public. People need to know how to protect their homes and families from fire. This might mean teaching children about fire safety and the dangers of playing with matches, or advice on the best methods of building houses and other structures so they can slow the progress of a fire. No building can be made completely **fireproof**—even materials such as concrete and brick can be destroyed in a large fire. But **building regulations** in many countries call for sprinkler systems, multiple ways out of a building, and non-flammable fire doors and firewalls to protect staircases and hallways leading to fire exits.

LINKING THE WORLD

The United Nations (U.N.) has become the most important organization for helping countries cooperate in emergencies. By helping smaller countries use technology and equipment developed in the U.S. and elsewhere, the United Nations helps people prepare for and cope with disasters. In March 2001, several U.N. departments joined with the Global Fire Monitoring Center (GFMC) to concentrate on battling fires around the world. Since it was founded in 1998, the GFMC has analyzed data from satellite photographs to determine where fire poses a threat and where it might have broken out. It then passes on this information to the country where the threat has been located. Through the U.N./GFMC program, countries can exchange information on how and where fires are likely to develop and how best to respond if fires do break out. Best of all, the U.N./GFMC provides the quickest way of alerting other countries (and getting help fast) if a major wildfire breaks out.

Left: The best way to save homes and communities from destructive fires is to prevent them in the first place.

PROTECTING ANIMALS

Animals are the victims of many wildfires, either because of the flames themselves or because their woodland habitats are destroyed. Livestock—the cattle, sheep, and horses kept by many rural farmers— also face terrible danger from wildfires. Unless farmers plan ahead, many of these animals will be left behind to die. Farmers in the American West, Australia, Argentina, and other areas must learn and practice the advice that fire services offer on how to protect their livestock.

They must know the best places where livestock can be herded during a fire to keep them safe. Just as importantly, farmers need to keep barns and corrals well away from trees and bushes that are likely to burn in a wildfire.

Below: Wild animals and livestock often become the helpless victims of wildfires that destroy farms, pasture areas, and natural animal habitats.

FIRENET

Sharing information about fires and how best to fight them is one of the most effective ways to provide a safer future. Recently, fire experts from around the world have been using the Internet to pass on such information. One of the most important organizations for this exchange of information is FireNet. Firefighting experts from around the world put up information and links about nearly everything to do with fires. For example, Internet users can find information about Ohio rural and community fire protection, Hornsby Bushfire Services in Australia, the Fire Science Center at the University of New Brunswick, and emergency services in Finland. Experts can share the latest software models of how fires start and grow, which can help them develop new ways to fight fires.

Glossary

amphibious able to land or drive both on ground and in water

arson the crime of deliberately starting a fire

brigades groups or teams of people organized to work together for a specific purpose, such as fighting a fire

building regulations rules that govern the construction of new buildings in order to ensure that they are safe

combustion a chemical reaction that produces heat and light

conducts allows something such as electricity to pass through easily

downwind in the path of the wind blowing from something such as a fire

evacuate to remove people from a dangerous area

extinguished put an end to or put out

fire blanket a heavy blanket made from nonflammable material, used to smother fires

firebreak a gap cut in a forest or other area of flammable material to block the spread of a fire

fireproof able to withstand fire

fire-resistant able to withstand fire for a certain period

flammable likely to catch fire easily

helitankers helicopters that can carry very heavy loads

hot spots the centers of fires where flames are burning most fiercely

hydrants large faucets for connecting fire hoses to a water supply; in many cities, hydrants are placed on every street corner in case of fire

insurance protection against loss or damage; a person pays a set amount to an insurance company each year, and in return, the insurance company pays for damage in the event of a fire or other accident

National Guard part of the U.S. Army that serves within a particular state, but can be called out by the national government in an emergency

oxygen a gas present in the air that is necessary for fire and other types of combustion to occur

pickaxes long-handled tools with heavy, pointed heads, used to smash or knock down objects

professional paid to do a job full-time

reaction an action that takes place when two or more substances are combined

smokejumpers firefighters who parachute deep into a wildfire

suffocate to die because air cannot reach the lungs

volunteers people who do something by choice and without being paid

wildfires fires in forests, prairies, swamps, or other areas where there are few houses

Further Information

Books

Coe, Andrew, and The New York City Fire Museum. *FDNY: An Illustrated History of the Fire Department of New York*. Corona, Calif.: Odyssey Books, 2003.

Gorrell, Gena Kinton. *Catching Fire: The Story of Firefighting*. New York: Tundra Books, 1999.

Yoder, Karen. *Fire Kids: The Adventures of Hose Company No. 2*. Trabuco Canyon, Calif.: Stoney Creek Press, 2002.

Web sites

The Survive Alive Village
www.survivealive.org

FireFighting.com
www.firefighting.com

FirefightingLinks.com
www.firefightinglinks.com

FireNet International
www.fire.org.uk

Index